EUPHORIA BEGAN TO DWINDLE

Context of Scientific Discoveries

DIPAN KUMAR DAS

SUDIP KUMAR DAS

To the curious minds who dare to ask
"why" and "how,"

To the tireless explorers who venture
into the unknown,

To the scientists, researchers, and
innovators who dedicate their lives to
the pursuit of knowledge,

To the teachers who inspire the next
generation of thinkers,

To the dreamers who envision a brighter future through science,

To the collaborative spirit that unites us all in the quest for understanding,

This exploration is dedicated to you. Your unwavering curiosity and boundless ambition continue to illuminate the path of discovery, reminding us that the pursuit of knowledge is a journey that knows no bounds.

May your insatiable thirst for understanding continue to push the boundaries of human knowledge, inspire innovation, and shape a better world for us all.

Foreword

In the vast expanse of human history, few endeavors have ignited the human imagination and reshaped our world as profoundly as the relentless pursuit of knowledge. From the earliest philosophers pondering the nature of the cosmos to modern scientists unraveling the mysteries of the quantum realm, the story of scientific exploration is a testament to the boundless curiosity of the human spirit.

"Euphoria Began To Dwindle" invites us on a journey through the intricate and dynamic landscape of scientific discovery. It is a journey that traverses the peaks of triumph and the valleys of challenge, illuminating the highs and lows of progress in the quest to understand our universe.

As we embark on this exploration, we are reminded that the pursuit of knowledge is not merely a profession; it is a calling—a calling to unlock the secrets of existence, to confront the complexities of our world, and to shape a better future for humanity. It is a calling that transcends disciplines, generations, and borders, uniting us in a shared endeavor to explore the unknown.

In the pages that follow, we will delve into the euphoria of groundbreaking discoveries, the spark of innovation that surges through laboratories, the highs and lows of scientific progress, the erosion of euphoria in the face of challenges, and the renewal of purpose and perseverance that fuels the scientific spirit.

This exploration is an ode to the scientists, researchers, and thinkers who have dedicated their lives to the pursuit of knowledge. It is a celebration of the enduring human curiosity that propels us forward and the resilience that allows us to overcome obstacles in our quest for understanding.

As we journey through the pages of "Euphoria Began To Dwindle," may we be inspired by the boundless curiosity that drives scientific exploration, and may we recognize that the pursuit of knowledge is a journey that shapes not only our understanding of the universe but also our shared destiny.

Welcome to this exploration—a voyage through the remarkable world of science, where questions become answers, mysteries are unveiled, and the pursuit of knowledge knows no end.

Preface

In the annals of human history, there exists a profound story—a story of relentless curiosity, boundless ambition, and unwavering dedication. It is the story of scientific exploration, an epic journey through the realms of the known and the unknown. "Euphoria Began To Dwindle" seeks to unravel this intricate tale, a tapestry woven from the threads of discovery, innovation, challenge, and renewal.

As we stand on the precipice of knowledge, it is essential to reflect on the path that brought us here. This exploration is not a linear narrative but a multifaceted mosaic—a tapestry that portrays the highs and lows of scientific progress. It is a journey through the corridors of laboratories, observatories, and libraries, where the greatest minds of our time have striven to unlock the secrets of the universe.

The pages that follow will invite you into the euphoria of groundbreaking discoveries, where the boundaries of human understanding expand. You will glimpse the spark of innovation that ignites the hearts and minds of scientists, pushing them to explore uncharted territories.

Yet, this journey is not without its challenges. As we delve into the exploration, we will confront the erosion of euphoria—a phase where progress faces setbacks, where questions multiply faster than answers, and where the pursuit of knowledge encounters obstacles and ethical dilemmas.

But amidst the challenges, there is a theme of resilience—a renewal of purpose and perseverance that drives the scientific spirit forward. This exploration will illuminate the unwavering commitment of scientists to address the most pressing issues of our time, from climate change to pandemics, from sustainable solutions to the mysteries of the cosmos.

In the preface of "Euphoria Began To Dwindle," we set the stage for this voyage—a journey through the wonders and complexities of scientific exploration. As we turn the pages, may you be inspired by the indomitable human curiosity that propels us to seek understanding, and may you come to appreciate that the pursuit of knowledge is a journey that shapes not only our understanding of the universe but also the very essence of humanity.

Welcome to this exploration—an odyssey through the realms of science, where the pursuit of knowledge knows no bounds and where the human spirit soars in the quest for understanding.

Prologue

In the vast expanse of the cosmos, on the tiny blue dot we call home, there exists a story—a story of human endeavor that spans millennia. It is a tale of insatiable curiosity, boundless ambition, and unwavering dedication. It is the story of scientific exploration, a journey through the mysteries of the universe that has shaped our understanding of the world and our place within it.

"Euphoria Began To Dwindle" embarks on a voyage through this remarkable narrative, a narrative that encapsulates the very essence of human exploration and discovery. It is a journey that takes us from the earliest philosophers gazing at the stars to the modern scientists unraveling the fabric of reality at the quantum level.

As we prepare to delve into this exploration, it is important to recognize that the pursuit of knowledge is not just a vocation; it is a calling—a calling to unravel the secrets of existence, to grapple with the complexities of our world, and to forge a path toward a brighter future for humanity. It is a calling that transcends generations, disciplines, and borders, uniting us in a shared endeavor to explore the unknown.

In the pages that follow, we will traverse the euphoria of groundbreaking discoveries, where the boundaries of human understanding expand with each revelation. We will witness the spark of innovation that ignites the hearts and minds of scientists, propelling them to explore uncharted territories.

Yet, this journey is not without its challenges. As we progress through the exploration, we will confront the erosion of euphoria—a phase where progress encounters setbacks, where questions multiply faster than answers, and where the pursuit of knowledge faces obstacles and ethical dilemmas.

But amid the challenges, there is a theme of resilience—a renewal of purpose and perseverance that propels the scientific spirit forward. This exploration will shine a light on the unwavering commitment of scientists to address the most pressing issues of our time, from climate change to pandemics, from sustainable solutions to the mysteries of the cosmos.

In the prologue of "Euphoria Began To Dwindle," we set the stage for this odyssey—a journey through the wonders and complexities of scientific exploration. As we turn the pages, may you be inspired by the indomitable human curiosity that propels us to seek understanding, and may you come to appreciate that the pursuit of knowledge is a journey that shapes not only our understanding of the universe but also the very essence of humanity.

Welcome to this exploration—a voyage through the realms of science, where questions become answers, mysteries are unveiled, and the pursuit of knowledge knows no end.

CHAPTER ONE

The Dawn of Euphoria

In the vast tapestry of human history, there have been moments when the pursuit of knowledge has ignited a brilliant euphoria, casting aside the shadows of ignorance and ushering in new eras of understanding. This chapter embarks on a journey through time, to an epoch when the world was on the cusp of scientific enlightenment, and the first rays of euphoria began to pierce the darkness of ignorance.

The dawn of euphoria is a time of intellectual awakening, a period when the human mind begins to fathom the intricacies of the natural world in ways previously unimaginable. It is an era marked by the courage to question, the audacity to challenge conventional wisdom, and the relentless pursuit of truth.

Our story unfolds during the Renaissance, a period of profound transformation in Europe. It is a time when the shackles of medieval dogma are being cast off, and a thirst for knowledge and discovery sweeps across the continent. In the dimly lit chambers of monasteries and the bustling workshops of artisans, seeds of curiosity are sown, and the first stirrings of a scientific revolution take root.

At the heart of this burgeoning euphoria are visionaries like Copernicus, who dares to challenge the geocentric view of the universe that has held sway for centuries. With trembling hands and an unwavering conviction, he pens "De

Revolutionibus Orbium Coelestium," positing that the Earth is not the center of the cosmos but rather one of many celestial bodies in orbit around the Sun.

In the universities of Europe, scholars laboriously dissect the works of ancient philosophers, seeking to decipher the laws that govern the natural world. Alchemy and astrology, once shrouded in mysticism, give way to the nascent disciplines of chemistry and astronomy. The pursuit of knowledge becomes a collective endeavor, as scholars from different lands exchange ideas, books, and manuscripts, setting the stage for a grand intellectual renaissance.

The dawn of euphoria also illuminates the lives of trailblazing women in science, like Hypatia of Alexandria and Maria Sibylla Merian, who defy societal norms to explore the mysteries of the natural world. Their contributions, though often overlooked in their time, serve as a testament to the indomitable human spirit that fuels the scientific quest.

As the first chapter unfolds, the world witnesses the birth pangs of a scientific revolution. The euphoria is palpable, like a brilliant sunrise after a long and arduous night. It is a time when humanity takes its first tentative steps towards unraveling the enigmatic tapestry of the cosmos, and in doing so, forever changes the course of history.

The euphoria of the Renaissance's scientific awakening extends far beyond the realm of astronomy and the heliocentric model. It touches every facet of human understanding, from the microcosmic world of cells to the macrocosmic expanses of the universe. As we delve deeper into this chapter, we encounter further examples of groundbreaking discoveries and pioneering individuals who epitomized the spirit of this epoch.

Vesalius and the Anatomy of Euphoria:

Andreas Vesalius, a Belgian physician, emerges as a pivotal figure during this era. His work, "De Humani Corporis Fabrica," represents a paradigm shift in the understanding of human anatomy. Vesalius challenges the long-held teachings of Galen, who relied on animal dissections, by conducting meticulous human cadaver dissections. His illustrations and detailed observations reveal the intricacies of the human body in unprecedented detail, shattering the dogmas of the past and ushering in a new era of medical knowledge.

The Ingenious Kepler:

Johannes Kepler, an astronomer and mathematician, follows in the footsteps of Copernicus. His laws of planetary motion, described in "Astronomia Nova," provide the mathematical framework that explains the elliptical orbits of planets around the Sun. Kepler's work not only confirms the heliocentric model but also demonstrates the power of mathematics as a tool for understanding the natural world.

The Age of Exploration:

The dawn of euphoria extends beyond Europe to the high seas. Explorers like Christopher Columbus, Ferdinand Magellan, and Vasco da Gama voyage to uncharted lands, expanding the known world and challenging the limits of navigation and cartography. These daring adventurers bring back new species, exotic plants, and previously unseen cultures, enriching the collective knowledge of humanity.

Galileo's Celestial Revelations:

Galileo Galilei, an Italian polymath, revolutionizes the world of astronomy with his telescopic observations. In "Sidereus Nuncius" and "Dialogue Concerning the Two Chief World Systems," he unveils the moons of Jupiter, the phases of Venus, and the rugged lunar surface, providing compelling evidence for Copernican heliocentrism. His work sparks not only scientific enthusiasm but also controversies with the Catholic Church.

Advancements in Art and Innovation:

The Renaissance is not solely about scientific breakthroughs. It is also a time of remarkable artistic expression and technological innovation. Leonardo da Vinci's notebooks are filled with designs and inventions that span a myriad of disciplines, from flying machines to anatomical studies. The blending of art and science during this period exemplifies the interdisciplinary nature of the Renaissance's euphoria.

The Alchemical Pursuit:

The Renaissance also witnesses the transition from alchemy, a mystical and proto-scientific discipline, to the more rigorous study of chemistry.

Alchemists like Paracelsus and Robert Boyle begin to lay the groundwork for modern chemistry, gradually shifting the focus from the transmutation of base metals into gold to the systematic investigation of matter, elements, and chemical reactions. The allure of unlocking the secrets of nature continues to fuel the euphoria of discovery.

The Printing Press Revolution:

A technological marvel of the Renaissance, the printing press, invented by Johannes Gutenberg, amplifies the dissemination of knowledge. With the rapid production of books and scientific treatises, ideas flow across Europe like never before, facilitating the exchange of insights and accelerating the pace of discovery.

The Unconventional Naturalists:

Naturalists such as Conrad Gessner and John Ray embark on expeditions to document and categorize the astonishing diversity of life on Earth. Their meticulous observations and classification systems lay the foundation for modern biology. The Renaissance's spirit of inquiry extends to the natural world, revealing an intricate web of species waiting to be understood.

The Multifaceted Genius, Copernicus:

Nicolaus Copernicus, often hailed as the father of modern astronomy, exemplifies the interdisciplinary nature of Renaissance thought. In addition to his heliocentric model, Copernicus is a polymath with interests in medicine, economics, and mathematics. His revolutionary ideas, grounded in both observation and mathematical reasoning, epitomize the Renaissance's holistic approach to understanding the world.

The Spirit of Collaboration:

Throughout this chapter, it becomes evident that the Renaissance is not defined by solitary geniuses alone but by collaborative networks of scholars, artists, and thinkers. Intellectual salons, academies, and universities become hubs of discussion and exchange, nurturing an environment where ideas flourish and the euphoria of discovery knows no bounds.

In the annals of scientific history, there have been pivotal moments when groundbreaking discoveries were made, igniting the initial euphoria that would fuel scientific exploration for years to come. These moments were like sparks in the darkness, setting ablaze the human curiosity and inspiring a generation of scientists and scholars. As we journey through this chapter, we will explore some of these turning points in the history of science.

The Enlightenment of Copernicus:

The year is 1543, and Nicolaus Copernicus, a Polish mathematician and astronomer, releases his magnum opus, "De Revolutionibus Orbium Coelestium." In this groundbreaking work, he presents the heliocentric model of the solar system, with the Sun at its center, challenging the geocentric view that had prevailed for centuries. Copernicus' work sparks a revolution in our understanding of the cosmos, setting the stage for future astronomical discoveries and the eventual acceptance of the heliocentric model.

Galileo's Celestial Revelations:

Galileo Galilei, an Italian polymath, takes his telescope to the night skies in the early 17th century. His observations of the Moon's surface, the moons of Jupiter, and the phases of Venus provide compelling evidence for the heliocentric model. In "Sidereus Nuncius" and "Dialogue Concerning the Two Chief World Systems," he challenges the geocentric worldview and ignites a scientific revolution. The euphoria generated by Galileo's discoveries reverberates through the corridors of academia and inspires generations of astronomers.

Newton's Universal Law of Gravitation:

In 1687, Isaac Newton publishes his monumental work, "Philosophiæ Naturalis Principia Mathematica" ("Mathematical Principles of Natural

Philosophy"). In this treatise, he formulates the laws of motion and the universal law of gravitation. Newton's equations provide a comprehensive framework for understanding the motion of celestial bodies and terrestrial objects. This discovery ushers in a new era of physics and engineering, fueling the scientific exploration of the natural world.

The Discovery of Electricity:

The 18th century witnesses the electrifying experiments of Benjamin Franklin and others. Franklin's famous kite experiment in 1752 provides crucial insights into the nature of electricity and lightning. These discoveries pave the way for the development of electrical theory and applications, revolutionizing technology and communication in the centuries that follow.

Darwin's Theory of Evolution:

In 1859, Charles Darwin publishes "On the Origin of Species," introducing the world to the concept of evolution through natural selection. This groundbreaking theory transforms biology and challenges prevailing ideas about the diversity of life on Earth. Darwin's work sparks a fervor of scientific investigation into the origins and evolution of species, setting the stage for modern evolutionary biology.

These moments in the history of science are just a glimpse of the countless discoveries that have ignited euphoria and driven the relentless pursuit of knowledge. Each breakthrough represented a departure from established dogma, a shattering of old paradigms, and an invitation to explore the mysteries of the universe with fresh eyes. The initial euphoria sparked by these discoveries would serve as a beacon, guiding generations of scientists toward new frontiers of understanding and innovation.

CHAPTER TWO

The Spark of Innovation

The world was abuzz with excitement, and the halls of academia and laboratories across the globe were filled with an electric energy. The scientific community, having basked in the glory of recent groundbreaking discoveries, now stood on the precipice of a new era. Chapter 2 of our journey through "Euphoria Began To Dwindle" takes us deep into this era, where the spark of innovation ignited a fervor for exploration and experimentation that would shape the course of history.

Uncharted Depths: The Oceanic Mysteries:

In the wake of the Age of Enlightenment, sailors and scientists embarked on daring voyages to explore the uncharted depths of the world's oceans. The HMS Challenger expedition of the 1870s, for instance, crisscrossed the global oceans, revealing a plethora of new species and geological wonders. Marine biology was born, and the oceans became a vast laboratory, sparking the imaginations of marine scientists who would unveil the secrets of the deep.

Masters of the Microcosm: Microbiology Emerges:

While astronomers gazed at the cosmos, another frontier was unfolding under the lens of microscopes. Scientists like Antonie van Leeuwenhoek and Louis Pasteur delved into the microscopic world, discovering bacteria, yeast, and other microorganisms. This new field of microbiology not only transformed our understanding of disease but also gave birth to the field of biotechnology, where microbes would be harnessed for myriad applications.

Electric Dreams: The Age of Electromagnetism:

The 19th century witnessed a dazzling array of discoveries in the field of electromagnetism. Michael Faraday's experiments with electric and magnetic fields laid the foundation for electric generators and motors. James Clerk Maxwell's equations elegantly described the relationship between electricity and magnetism, leading to the prediction of electromagnetic waves, which we now know as light. These innovations sparked the electrification of society, changing the way people lived and worked.

The Space Race: A Cosmic Odyssey:

In the mid-20th century, the spark of innovation catapulted humanity into the cosmos. The Space Race between the United States and the Soviet Union resulted in the first human in space, Yuri Gagarin, and the historic Apollo 11 moon landing led by Neil Armstrong. These audacious missions marked the pinnacle of human exploration and sparked a new era of space science and technology.

Genetics Unveiled: The Double Helix:

In 1953, James Watson and Francis Crick unlocked the structure of DNA, revealing the elegant double helix. This discovery not only transformed biology but also ignited the field of genetics. Scientists could now unravel the genetic code, paving the way for groundbreaking developments in medicine, biotechnology, and genetic engineering.

Chemical Revolution: The Periodic Table Unveiled:

Dmitri Mendeleev's creation of the periodic table in 1869 was a monumental leap in chemistry. By arranging the known elements into a systematic framework based on atomic number and properties, Mendeleev not only revealed hidden patterns but also predicted the existence of yet-to-be-discovered elements. This innovation provided a roadmap for understanding matter at its fundamental level and fueled advances in chemistry, materials science, and industry.

The Quantum Leap: Quantum Mechanics and Quantum Theory:

The early 20th century ushered in a revolution in physics with the advent of quantum mechanics. Pioneers like Max Planck, Albert Einstein, Niels Bohr, and Werner Heisenberg challenged classical physics and introduced a new paradigm. Quantum mechanics illuminated the behavior of particles at atomic and subatomic scales, leading to the development of quantum theory. This quantum revolution paved the way for technologies such as semiconductors, lasers, and quantum computing, while fundamentally reshaping our understanding of the universe.

Environmental Awakening: Earth Sciences and Ecology:

The mid-20th century witnessed a growing awareness of the environment's fragility. Scientists like Rachel Carson, through her seminal work "Silent Spring," raised concerns about the impact of pesticides on ecosystems. This catalyzed the birth of modern ecology and environmental science, sparking a global movement for environmental conservation, sustainability, and the study of Earth systems.

Biotechnological Renaissance: The Dawn of Genetic Engineering:

The discovery of the structure of DNA laid the groundwork for genetic engineering. In the 1970s, scientists like Paul Berg and Herbert Boyer made history by successfully splicing DNA from different organisms, launching the biotechnological revolution. This innovation led to the development of genetically modified organisms (GMOs), gene therapy, and the decoding of entire genomes, including the Human Genome Project. The implications of these breakthroughs continue to shape medicine, agriculture, and ethics.

Cybernetic Frontier: The Information Age:

The late 20th century marked the advent of the Information Age. Innovations in computing, telecommunications, and the internet transformed the way humans communicate, work, and access knowledge. Pioneers like Tim Berners-Lee, the inventor of the World Wide Web, and Steve Jobs, the co-founder of Apple, ushered in an era where information became more accessible than ever before, sparking an ongoing digital revolution.

Nanotechnology: Engineering at the Molecular Scale:

The late 20th century and early 21st century saw the emergence of nanotechnology, a field that explores manipulating matter at the nanoscale. Scientists and engineers, inspired by the concept that "smaller is different," harnessed the unique properties of materials at the nanoscale. Innovations like carbon nanotubes, quantum dots, and nanomedicine opened new vistas in electronics, materials science, and healthcare, fueling a wave of research and development with profound implications for the future.

Renewable Energy Revolution:

Faced with environmental challenges and the finite nature of fossil fuels, the scientific community embarked on a quest for sustainable energy sources. Innovations in solar cells, wind turbines, and energy storage technologies transformed the energy landscape. The pursuit of clean, renewable energy sources became a global imperative, sparking innovation in energy production, storage, and distribution.

Artificial Intelligence and Machine Learning:

The 21st century has witnessed a remarkable resurgence of interest in artificial intelligence (AI) and machine learning (ML). Driven by exponential advances in computing power and data availability, researchers have developed AI systems that can analyze vast datasets, recognize patterns, and make predictions. AI innovations have revolutionized industries such as healthcare, finance, and transportation, sparking an ongoing revolution in automation and decision support.

Biomedical Breakthroughs: CRISPR and Beyond:

In recent years, the field of genetics has witnessed a revolution with the advent of CRISPR-Cas9 gene-editing technology. This breakthrough allows scientists to precisely edit genes, offering the potential to treat genetic diseases and transform agriculture. The excitement generated by CRISPR has led to a flurry of research and innovation in genetic medicine and biotechnology, promising a future where genetic diseases may be conquered.

Exploration Beyond Earth: Mars and Exoplanets:

The spark of innovation extends beyond our planet as humanity looks to the stars. Ambitious missions to Mars, the discovery of thousands of exoplanets, and the search for extraterrestrial life have ignited a renewed interest in space exploration. These endeavors, backed by advances in rocket technology and astrobiology, have sparked an era of renewed optimism in the quest to explore the cosmos.

Climate Science and Sustainability:

With the growing recognition of the Earth's changing climate and environmental challenges, the scientific community rallied to understand and address these issues. Innovations in climate science, including advanced climate modeling and satellite technology, have provided vital insights into the impacts of climate change. This has spurred global efforts to mitigate its effects and transition toward sustainable practices, igniting a revolution in clean energy, conservation, and environmental stewardship.

Neuroscience and Brain-Computer Interfaces:

The study of the human brain has experienced a renaissance in recent years. Neuroscientists are deciphering the brain's complex workings, unveiling its mysteries, and developing technologies such as brain-computer interfaces (BCIs). BCIs have the potential to revolutionize healthcare, allowing individuals with paralysis to control devices, and advancing our understanding of cognition and consciousness.

Quantum Computing:

Quantum computing represents a paradigm shift in information processing. Scientists have harnessed the peculiar properties of quantum bits (qubits) to perform calculations exponentially faster than classical computers. This innovation holds the promise of solving complex problems in fields such as cryptography, materials science, and drug discovery. The race to build practical quantum computers has ignited a fervor of research and development.

Genomic Medicine and Personalized Healthcare:

The decoding of the human genome and advancements in genomic sequencing technology have paved the way for personalized medicine. Innovations in genomics allow healthcare providers to tailor treatments to individual patients' genetic profiles, ushering in an era of precision medicine. This personalized approach holds the potential to revolutionize healthcare by optimizing treatments and reducing adverse effects.

Artificial Photosynthesis and Sustainable Agriculture:

Inspired by nature's ability to convert sunlight into energy, scientists are developing artificial photosynthesis technologies. These innovations aim to harness solar energy to produce fuels, chemicals, and food sustainably. Simultaneously, advancements in agriculture, including precision farming and genetically modified crops, are sparking a revolution in food production to meet the demands of a growing global population.

CHAPTER THREE

The Highs and Lows of Progress

In the relentless quest for knowledge, scientific progress is a journey fraught with peaks of exhilaration and valleys of frustration. Chapter 3 of our exploration, "The Highs and Lows of Progress," takes us through the rollercoaster ride of advancement and setbacks that scientists encountered as they sought to unravel the mysteries of the universe. It is a chapter that reveals the profound resilience of the human spirit and the unwavering commitment to the pursuit of understanding.

The Elusive Aether and Michelson-Morley Experiment:

The late 19th century was a time of great optimism in physics, yet it was marked by a puzzling problem: the search for the luminiferous aether. Scientists believed this mysterious substance was the medium through which light waves traveled. However, the Michelson-Morley experiment in 1887 failed to detect the aether, challenging fundamental assumptions about the nature of light and motion. This setback ultimately paved the way for Albert Einstein's theory of special relativity, a groundbreaking high point in physics.

The Quest for the Ether and the Collapse of the Phlogiston Theory:

Chemistry, too, faced moments of frustration. The once-prominent phlogiston theory, which posited the existence of an undetectable substance called phlogiston responsible for combustion, began to crumble under scrutiny in the late 18th century. While this was a low point in chemistry's history, it led to the development of the modern understanding of oxidation and the birth of the science of chemistry as we know it today.

Eclipses and Einstein's Triumph:

In 1915, Albert Einstein proposed his theory of general relativity, which predicted that gravity could bend the path of light. The concept was put to the test during a solar eclipse in 1919 when scientists observed stars whose light was bent by the sun's gravitational field. This experimental confirmation of Einstein's theory marked a remarkable high point in the history of physics and demonstrated the power of theoretical predictions.

The Challenges of Space Exploration:

Space exploration has seen both awe-inspiring achievements and heart-wrenching setbacks. The early days of spaceflight witnessed triumphs like Yuri Gagarin's historic orbit of the Earth in 1961. However, it was also marred by tragedies, including the Apollo 1 fire and the Challenger and Columbia disasters. These losses were painful reminders of the risks inherent in pushing the boundaries of exploration, but they did not deter humanity's drive to explore the cosmos.

Genetic Engineering and Ethical Quandaries:

The advent of genetic engineering and CRISPR-Cas9 technology in recent decades has ushered in a new era of genetic manipulation. While the potential to cure genetic diseases and enhance human capabilities is tantalizing, it also raises profound ethical and moral questions. The scientific community grapples with these challenges, marking a period of both excitement and ethical introspection.

Climate Science and the Urgency of Action:

The field of climate science has faced its own set of highs and lows. While the accumulation of evidence pointing to human-induced climate change is a scientific triumph, the slow pace of global action to mitigate its effects remains a source of frustration. Scientists find themselves at the intersection of science, policy, and public perception, struggling to convey the urgency of addressing climate change.

The Quantum Quandary:

Quantum mechanics, despite its profound success, has been a source of both wonder and vexation for scientists. Its bizarre principles, like entanglement and superposition, challenge our intuitive understanding of the physical world. The famous Schrödinger's cat paradox encapsulates this enigmatic aspect of quantum theory. While quantum mechanics has enabled remarkable technological advancements, including quantum computing and cryptography, it has also left scientists grappling with philosophical and interpretational questions.

The Higgs Boson Hunt:

The search for the Higgs boson, often dubbed the "God particle," spanned decades and involved immense collaboration among scientists and colossal particle accelerators like the Large Hadron Collider (LHC). The discovery of the Higgs boson in 2012 validated the Standard Model of particle physics, a monumental achievement. Yet, it also raised new questions, such as the nature of dark matter and the limitations of the Standard Model, underscoring the perpetual quest for deeper insights.

The Promise and Perils of Artificial Intelligence:

While artificial intelligence (AI) has made astonishing strides, it has also provoked concerns about its societal impact. The rise of AI-powered autonomous systems and machine learning algorithms has given rise to ethical dilemmas, including bias in algorithms, job displacement, and concerns about the loss of human control. The scientific community grapples with the responsibility of guiding AI's development to benefit humanity.

Medical Miracles and Ethical Boundaries:

Advances in medical science and biotechnology have achieved remarkable breakthroughs, from organ transplantation to stem cell therapies. However, these medical miracles are not without ethical and moral complexities. Debates over issues such as cloning, genetic editing, and the creation of designer babies have raised profound questions about the limits of scientific progress and the sanctity of life.

Planetary Exploration and the Mysteries of Mars:

Mars, the "Red Planet," has long fascinated scientists with the possibility of past or present life. While missions like the Mars rovers and landers have provided tantalizing clues about the planet's history, they have yet to definitively answer the question of Martian life. The quest for extraterrestrial life remains both a scientific aspiration and a humbling reminder of the vastness of the cosmos.

The Power and Responsibility of Gene Editing:

CRISPR-Cas9 gene-editing technology offers unprecedented

power to modify genes and potentially eradicate genetic diseases. However, it also presents ethical dilemmas, such as the unintended consequences of gene editing, and the potential for designer babies. Scientists and ethicists grapple with how to navigate the fine line between progress and responsibility in the realm of genetic engineering.

CHAPTER FOUR
The Erosion of Euphoria

As the sands of time continued to flow, the once radiant euphoria that had enveloped the scientific community began to erode. In Chapter 4, we delve into the intricate tapestry of this transformation, where the initial jubilation that accompanied groundbreaking discoveries slowly gave way to a sense of disillusionment. The journey of scientific exploration, once characterized by boundless optimism, now navigated the complexities of progress, unveiling the challenges that would test the resilience of scientific endeavor.

The Diminishing Low-Hanging Fruit:

In the early days of scientific discovery, it seemed that every experiment and observation revealed new and astonishing insights. The thrill of uncovering the low-hanging fruit of knowledge was intoxicating. However, as time passed, these easy wins became increasingly scarce, and scientists found themselves confronted with ever more complex and elusive questions. The initial sense of euphoria, once sustained by rapid progress, began to wane.

The Struggle for Funding and Resources:

The erosion of euphoria was further exacerbated by the increasingly competitive landscape of scientific research. As funding became scarcer and research budgets tighter, scientists faced fierce competition for resources. Grant applications, once a mere formality, became protracted battles, with many promising projects going unfunded. The constant struggle for financial support added a layer of stress and uncertainty that weighed heavily on the scientific community.

The Ethical Quandaries of Innovation:

As scientific discoveries pushed the boundaries of what was possible, they also raised profound ethical dilemmas. Innovations in genetics, artificial intelligence, and biotechnology posed questions about the responsible use of knowledge. The realization that scientific progress could have unintended consequences led to debates about the ethical implications of research and the need for responsible stewardship.

The Reproducibility Crisis:

The erosion of euphoria was exacerbated by the so-called reproducibility crisis. As scientific rigor came under scrutiny, it became evident that not all published research could be reliably replicated. This crisis cast doubt on the validity of some scientific findings and led to a call for more robust experimental design and transparency in research practices.

The Paradox of Information Overload:

The Information Age brought with it an overwhelming flood of data and publications. While the accessibility of information was a boon, it also became a burden. Scientists found themselves grappling with the paradox of information overload, where the sheer volume of research made it increasingly challenging to discern valuable insights from noise.

The Increasing Complexity of Scientific Questions:

As scientific knowledge expanded, so did the complexity of the questions being asked. Researchers delved into the intricacies of fields such as quantum mechanics, string theory, and neuroscience, where the mysteries ran deep. The pursuit of answers in these complex realms often required decades of dedication and collaboration, testing the patience and resolve of scientists.

The Hurdles of Interdisciplinary Research:

As scientific questions became increasingly multifaceted, the need for interdisciplinary collaboration grew. While interdisciplinary research held great promise, it also posed challenges in terms of communication, integration of knowledge, and the negotiation of differing research paradigms. This complexity added layers of difficulty to scientific inquiry, sometimes leading to frustration.

The Burden of Bureaucracy and Red Tape:

Scientific research, once characterized by curiosity-driven exploration, began to grapple with the growing burden of bureaucracy. The demands of compliance, paperwork, and regulations escalated, diverting valuable time and resources away from scientific inquiry. This administrative burden weighed on scientists, who yearned for a return to the unbridled pursuit of knowledge.

The Impact of Public Skepticism and Misinformation:

The erosion of euphoria was also influenced by a rise in public skepticism towards science and the spread of misinformation. The advent of the internet and social media enabled the rapid dissemination of both accurate and misleading information, eroding public trust in scientific expertise. Scientists found themselves not only communicating their discoveries but also defending the very foundations of scientific inquiry.

The Challenge of Global Crises:

The 21st century brought forth a series of global crises, including pandemics, climate change, and geopolitical tensions. These challenges demanded the attention and expertise of the scientific community. While scientists rose to the occasion, the weight of these crises added additional layers of complexity and urgency to their work, testing their resilience and adaptability.

The Necessity of Science Communication and Outreach:

The erosion of euphoria underscored the importance of effective science communication and outreach.

Scientists began to recognize that their role extended beyond the laboratory or fieldwork; they needed to engage with the public, policymakers, and stakeholders to convey the value of science and its role in addressing pressing global challenges.

The Quest for Revival and Rejuvenation:

Despite the challenges and complexities, scientists remained committed to their quest for knowledge. Many sought ways to rekindle the sense of euphoria and wonder that had initially drawn them to their fields. They embraced new approaches to collaboration, transparency, and communication, striving to reignite the excitement of discovery that had once fueled their pursuits.

The Burden of Commercialization and Patents:

In the modern era, the process of turning scientific discoveries into commercial products often strained the once-pure pursuit of knowledge. The need to secure patents, navigate intellectual property rights, and balance the demands of profit and public good introduced a layer of complexity that sometimes conflicted with the traditional ideals of scientific research.

The Human Cost of Scientific Endeavor:

The erosion of euphoria also cast a spotlight on the human toll of scientific endeavors. The relentless pursuit of knowledge often came at personal sacrifices, including long hours, stress, and strained work-life balances. Scientists grappled with the impact on their physical and mental well-being, prompting discussions about the need for greater work-life balance and mental health support in scientific communities.

The Challenge of Inclusivity and Diversity:

Science grappled with issues of inclusivity and diversity as underrepresented groups fought for recognition and opportunities in scientific fields. The erosion of euphoria underscored the need to address systemic biases and create more inclusive environments where the full spectrum of human talent could contribute to scientific progress.

The Ambiguity of Scientific Truth:

The pursuit of scientific truth sometimes confronted the ambiguity inherent in complex systems and phenomena. In fields like climate science, where predicting the future relies on intricate models and vast datasets, communicating uncertainty became a central challenge. The erosion of euphoria highlighted the necessity of effectively conveying the nuanced nature of scientific knowledge to the public and policymakers.

The Resurgence of Science Advocacy:

While the erosion of euphoria brought challenges, it also sparked a resurgence of science advocacy. Scientists, recognizing the importance of their work in addressing global challenges, became more active in advocating for science-based policies and greater investments in research and education. This renewed engagement sought to rebuild public trust in science and inspire future generations of scientists.

CHAPTER FIVE
Renewed Purpose and Perseverance

In the final chapter of our exploration, we enter a phase of renewal and resilience within the scientific community. As the initial euphoria waned, scientists did not lose faith in the power of discovery; instead, they redirected their focus toward addressing the pressing challenges of our time. Chapter 5, "Renewed Purpose and Perseverance," sheds light on the enduring spirit of scientific inquiry and how, despite the dwindling euphoria, the pursuit of knowledge remains an indomitable force in an ever-evolving world.

The Call to Action on Climate Change:

The erosion of euphoria compelled scientists to confront the looming crisis of climate change with renewed vigor. They embraced the role of climate advocates and champions, sounding the alarm about the consequences of inaction. Scientific research became the foundation for global climate agreements, spurring international cooperation and commitments to reduce greenhouse gas emissions.

Pandemics and the Pursuit of Solutions:

The emergence of global pandemics, like the COVID-19 crisis, underscored the critical role of science in addressing immediate and far-reaching challenges. Scientists mobilized to develop vaccines, treatments, and public health strategies, demonstrating the power of scientific collaboration and innovation in the face of a crisis.

The Resurgence of Space Exploration:

Space exploration experienced a renaissance, driven by renewed purpose and international cooperation. Missions to Mars, the search for extraterrestrial life, and plans for lunar habitats captured the imagination of scientists and the public alike. The pursuit of knowledge beyond Earth's boundaries once again ignited the flames of curiosity and ambition.

The Ethical Imperative in Research:

Scientists recognized the ethical imperative embedded in their work. They sought to navigate the complexities of innovation with a heightened awareness of the potential consequences. Ethical considerations became an integral part of the scientific process, guiding decisions and actions to ensure the responsible use of knowledge.

Science Communication and Public Engagement:

Scientists redoubled their efforts in science communication and public engagement. They recognized that effectively conveying the value of science and its role in addressing societal challenges was crucial to building public trust and support. The erosion of euphoria prompted scientists to become not just discoverers but also ambassadors of knowledge.

Diversity, Equity, and Inclusion in Science:

The scientific community embraced a commitment to diversity, equity, and inclusion. Efforts were made to dismantle systemic barriers and create an inclusive environment where talent from all backgrounds could flourish. This renewal of purpose aimed to ensure that science truly represented the diversity of human perspectives and experiences.

The Enduring Quest for Knowledge:

Above all, the erosion of euphoria reaffirmed the enduring quest for knowledge. Scientists, no longer solely driven by the thrill of discovery, found their purpose in addressing the world's most pressing challenges. They recognized that the pursuit of understanding was not a fleeting endeavor but a lifelong commitment to advancing humanity's collective knowledge.

Global Collaborations for Scientific Progress:

The erosion of euphoria led to a greater emphasis on global collaborations in scientific research. Scientists recognized that solving complex, global challenges required the collective wisdom and resources of the entire international community. Projects like the International Space Station, the Large Hadron Collider, and collaborative efforts in climate science became symbols of unity in the pursuit of knowledge.

Scientific Advances in Renewable Energy:

In response to environmental concerns, scientists redoubled their efforts to advance renewable energy technologies. Solar panels, wind turbines, and energy storage systems saw remarkable progress. The quest for sustainable energy sources became a central focus, offering hope for a greener, more sustainable future.

AI and Machine Learning for the Benefit of Humanity:

Despite concerns about the impact of artificial intelligence, scientists worked tirelessly to harness its potential for the betterment of humanity. AI and machine learning found applications in healthcare, disaster prediction, and optimizing resource management. The focus shifted from the fear of job displacement to the promise of increased efficiency and improved decision-making.

Empowering the Next Generation of Scientists:

The erosion of euphoria prompted a renewed commitment to educating and inspiring the next generation of scientists. Mentorship programs, science outreach initiatives, and educational reforms aimed to equip young minds with the tools and passion needed to tackle the world's most complex challenges.

Adaptation and Resilience in the Face of Uncertainty:

As the world grappled with uncertainty, the scientific community demonstrated remarkable adaptability and resilience. Researchers in fields like epidemiology and climate science adapted their models and strategies to address evolving threats. The erosion of euphoria underscored the need for science to remain agile in the face of rapid change.

The Rekindling of Wonder and Awe:

While the initial euphoria may have dimmed, scientists found that wonder and awe could still be kindled in their work. They discovered that the pursuit of knowledge, even in the face of challenges, continued to reveal the astonishing complexity and beauty of the universe. Each new discovery, no matter how incremental, reignited the flames of curiosity and wonder.

The Quest for Sustainable Solutions:

Scientists embraced the challenge of finding sustainable solutions to the world's most pressing problems. They explored innovative approaches to food production, transportation, and urban planning, seeking to reduce the human impact on the environment. The pursuit of sustainability became a driving force, balancing progress with the need to protect our planet for future generations.

Resilience in the Face of Adversity:

The erosion of euphoria tested the resilience of the scientific community. In the wake of setbacks and disappointments, scientists demonstrated remarkable fortitude. They viewed failures not as obstacles but as opportunities to learn and refine their approach. This resilience allowed them to persevere in the pursuit of answers to complex questions.

The Interconnectedness of Scientific Disciplines:

Scientists recognized the interconnectedness of different scientific disciplines. They sought to bridge gaps between fields, realizing that solutions to complex challenges often required a holistic approach. Interdisciplinary collaboration became a hallmark of scientific progress, leading to breakthroughs that transcended traditional boundaries.

The Impact of Citizen Science and Public Engagement:

The erosion of euphoria saw an increased emphasis on citizen science and public engagement. Scientists actively involved the public in research projects, recognizing that the collective intelligence of communities could contribute valuable insights. This approach fostered a sense of shared ownership in the pursuit of knowledge.

The Enduring Human Spirit of Curiosity:

Ultimately, the erosion of euphoria reaffirmed the enduring human spirit of curiosity. Scientists and citizens alike continued to ask questions, seek answers, and push the boundaries of knowledge. The pursuit of understanding remained an intrinsic part of the human experience, inspiring individuals to explore the world and the universe.

Epilogue

In the grand tapestry of human history, the story of scientific exploration is a testament to the unyielding human spirit. "Euphoria Began To Dwindle" has taken us on a journey through the triumphs and tribulations of scientific progress, from the initial spark of discovery to the erosion of euphoria and the renewal of purpose and perseverance.

As we reflect on this journey, we find that the pursuit of knowledge is not a linear path but a dynamic and evolving one. It is a journey characterized by the relentless curiosity of the human mind, the boundless ambition of scientists, and the unwavering commitment to uncovering the mysteries of the universe.

Throughout our exploration, we have witnessed how the scientific community adapts to the changing tides of progress, how it confronts challenges and setbacks with resilience, and how it rediscovers its sense of purpose in addressing the most pressing issues of our time. The erosion of euphoria, far from being a signal of defeat, is a testament to the enduring strength of scientific inquiry.

As we conclude our journey, we are reminded that the pursuit of knowledge is a shared endeavor, one that transcends borders, disciplines, and generations. It is a testament to the power of human collaboration, the humility to acknowledge what we do not yet know, and the determination to uncover the truth.

In the pages of this exploration, we have celebrated the highs and lows of scientific discovery, the complexities of progress, and the ethical dilemmas that arise along the way. We have witnessed the resilience of the human spirit and the enduring commitment to understanding our world and the universe beyond.

The story of "Euphoria Began To Dwindle" is not an end but a continuation—a call to inspire future generations to embark on their own journeys of exploration and discovery. It is a reminder that the pursuit of knowledge is a never-ending quest, one that will continue to shape our understanding of the universe and our place within it.

As we close this chapter of exploration, we do so with the knowledge that the pursuit of knowledge will forever be a guiding star in the vast expanse of human endeavor. It is a journey that will continue to inspire wonder, drive innovation, and light the path toward a future filled with new questions, new answers, and new horizons to explore.

…..***…..